Simple Geological Mapwork

An Introduction to Problem Maps
Without Strike Lines

W. E. Johnson

Edward Arnold

© W. E. Johnson 1976

First published 1976
by Edward Arnold (Publishers) Limited,
25 Hill Street, London W1X 8LL

ISBN 0 7131 0023 0

Filmset by Photoprint Plates Limited, Rayleigh, Essex
Printed in Great Britain by Butler & Tanner Ltd.,
Frome and London

Preface

This mapwork book is intended to provide an introduction to the main principles of geological mapwork as required for 'O' level and CSE courses, in which reference to strike lines is not required. No previous geological knowledge is assumed and although this book is concerned mainly with the interpretation of outcrops on geological maps, all the structural terms used are explained before their relationship to surface outcrops is illustrated. Exercises as an integral part of the text ensure the students' understanding of each topic as it is introduced.

Acknowledgements

The author is grateful to the following examination boards for permission to reproduce mapping questions from their CSE and 'O'-level papers: Map 1, 2 and 3, North Western Secondary School Examinations Board (CSE Summer 1973; CSE Summer 1974; CSE Summer 1974); Map 4, Joint Matriculation Board, ('O' level November 1973); Map 5, Oxford Local Examinations ('O' level, Autumn 1973).

1976   W.E.J.

Contents

Key to the main symbols used in this book

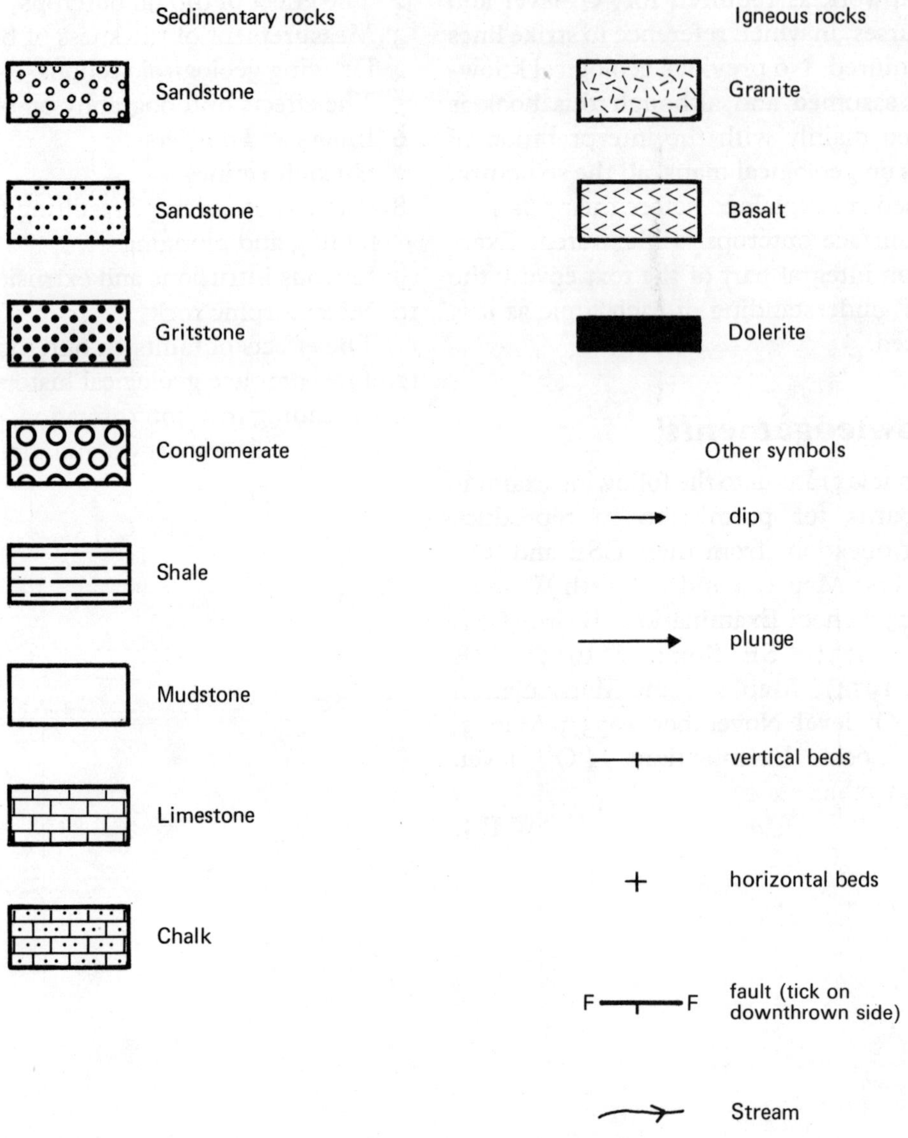

Sedimentary rocks

Sandstone

Sandstone

Gritstone

Conglomerate

Shale

Mudstone

Limestone

Chalk

Igneous rocks

Granite

Basalt

Dolerite

Other symbols

→ dip

⟶ plunge

—|— vertical beds

+ horizontal beds

F——|——F   fault (tick on downthrown side)

⟶ Stream

(other rock types are indicated in a key with the relevant diagram)

1 Some important terms and symbols

A geological map shows the position of the various types of rocks at the earth's surface. These rocks are in fact often covered by soil or even buildings, but it is the position of the rocks beneath which is mapped. The actual geological maps published by the Geological Survey are both colourful and complex, but on the problem maps in this book symbols are used for the different types of rocks and the maps are much simpler although the principles used in interpreting them are the same.

The part of the rock which appears at the surface (and is therefore shown on the map) is called *the outcrop* and a rock may be said *to outcrop*, e.g. in Fig. 1 there is a large outcrop of sandstone in the north west of the map and this sandstone also outcrops again in the east. If no indication of direction is given north should be taken as being at the top of the map.

Sedimentary rocks are deposited in layers, one on top of another to form *beds* or *strata*. They are then said to be *conformably bedded*. These strata are often visible in quarries and deep road cuttings. The division between two such layers or beds of rock is known as the *bedding plane*. The beds would originally have been deposited horizontally with the oldest (first formed) beds at the bottom, but they may later have been tilted so that they are said to *dip* (Fig. 2).

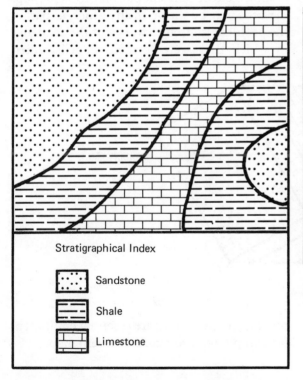

Stratigraphical Index

Sandstone

Shale

Limestone

Fig. 1 A simple geological map showing rock outcrops. The stratigraphical index shows the oldest rocks at the base of the column.

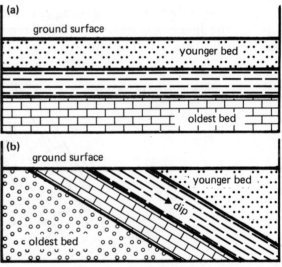

(a) ground surface — younger bed — oldest bed

(b) ground surface — younger bed — dip — oldest bed

Fig. 2 The two diagrams show vertical sections through the rocks beneath the ground.
(a) Horizontal beds.   **(b)** Dipping beds.

The *angle of dip* is the angle of the bed to the horizontal (Fig. 3). On a map this is shown by an arrow with the number of degrees.

Particular types of shading are commonly used to represent the various types of rock—three have already been used for sandstone, shale

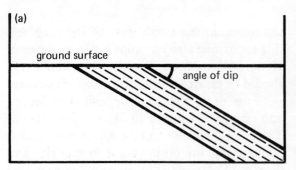

Fig. 3 **(a)** Section showing angle of dip. **(b)** Map showing dip arrows.

The symbol for horizontal beds is + and the symbol for vertical beds (ones which have been tilted through 90°) is ——+—— with the long line in the direction of the strike.

The *strike* of the rock is at right angles to the dip, i.e. the direction in which the beds extend horizontally (Fig. 4). The dip *(true dip)* is

and limestone. A complete key to the symbols used in this book is given opposite page 1.

In the key to a geological map the oldest rock is shown at the bottom, but on problem maps the beds may be shown in any order, as part of the problem is often to work out the sequence in which they were formed. Pencils, ruler,

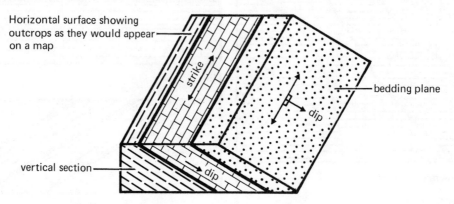

Fig. 4 Block diagram showing dip and strike.

therefore the steepest slope of the bed and can be seen in a section drawn parallel to the dip (at right angles to the strike). If the section is not drawn along the true dip but obliquely to it only the *apparent dip* will be seen and this will always be less than the true dip (Fig. 5).

protractor and rubber are essential equipment for completing geological problem maps.

2

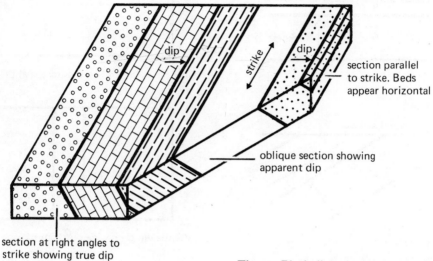

dip

strike

dip

section parallel
to strike. Beds
appear horizontal

oblique section showing
apparent dip

section at right angles to
strike showing true dip

Fig. 5 Block diagram showing the difference between true dip and apparent dip.

2 The effect of dip on outcrops

Exercise 1

Throughout this exercise the ground surface is taken as horizontal. The same series of beds A, B and C is used where A is the oldest and C is the youngest, and each bed is drawn with the same thickness in all four sections. X—Y shows the line of section. The first two sections and maps should be studied. Copy Maps (iii) and (iv) into your book; complete the outcrops, and the statements below each map.

SECTIONS	MAPS

(i) HORINZONTAL BED

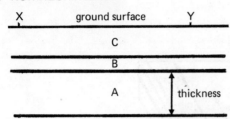

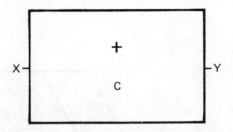

Only the youngest bed C outcrops at the surface

(ii) GENTLE DIP

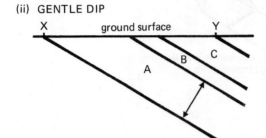

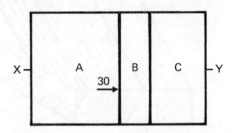

Gentle dip gives wide outcrops of each bed

(iii) STEEP DIP

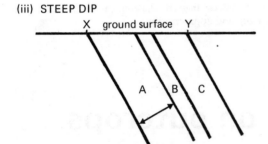

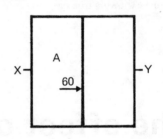

Steep dip gives _____ outcrops of each bed

(iv) VERTICAL BEDS

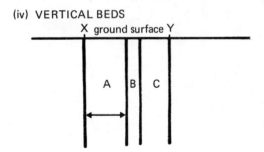

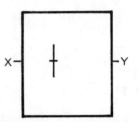

Vertical beds give very narrow outcrops
equal to the_____ of the beds

Three important rules can be deduced from the exercise above:

a The steeper the dip the narrower the outcrop (for a given thickness of bed).

b For a given angle of dip thicker beds have wider outcrops.

c The direction of dip is towards the younger beds (i.e. from A to C in above diagrams (ii) and (iii).

3 Measurement of thickness of beds

Where the angle of dip and scale of a map are known the thickness of the beds can be found. Two different measurements can be made on an accurately drawn section:

a The actual or *true thickness* (as referred to in Exercise 1).

b The *vertical thickness* (as would be found in a bore hole, Fig. 6).

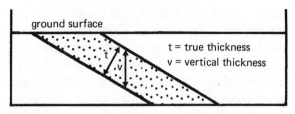

Fig. 6 Section showing true thickness and vertical thickness.

4 Drawing a geological section

First draw an outline section to show the relief of the ground. Where problem maps have no contours this is normally provided. The vertical scale of the section should be the same as the horizontal scale of the map, so that both the angle of dip and the thickness of the beds can be drawn in precisely. If the scales are not the same it is not possible to draw both correctly and therefore the thickness cannot be accurately measured.

Sections can be drawn in any direction across a map, but are best made in the direction of true dip, as it is then possible to draw in the angle of true dip accurately with a protractor. In many cases the line of section is given. If the section is not in the direction of true dip then the apparent dip, at a smaller angle, should be shown. If the section is nearly along the line of true dip the angle shown should be nearly as great as the true angle of dip, but if the section

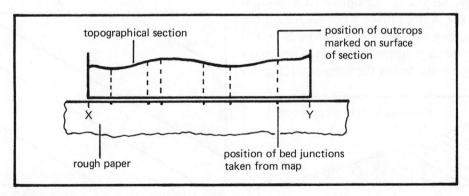

Fig. 7 How to mark in the position of rock outcrops on a topographical section.

is nearly along the strike the apparent dip is very slight. Where a section is drawn exactly along the strike the beds will appear horizontal (Fig. 5).

To draw in the geology beneath the outline section mark the position of the outcrops *on the surface* of the section. This can be done in the following way: lay the edge of a piece of rough paper across the map, along the line of section. Mark with a pencil where the different beds cross the paper. Then place the paper exactly beneath your outline section and mark the beds on the ground surface, *vertically above* their position on the rough paper (see Fig. 7).

Do the same for all the bed junctions and join up the beds. Where the thickness of the oldest bed cannot be found, a dashed line should be drawn to show a possible position of the base of the bed. If a bed outcrops in two or more places and its dip is the same all over the map (uniform dip), it is possible to draw it in by joining up the points marking the top and the points marking the bottom of the bed (this is best done where there are three or more points to join up). Accuracy is essential when drawing a section. Shading should be drawn to run in the direction of bedding and the beds may also be coloured in (Fig. 9).

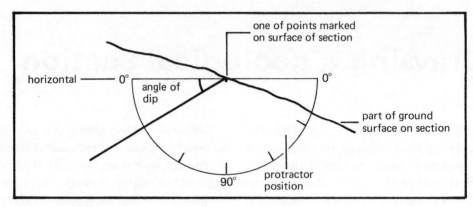

Fig. 8   How to draw in beds of rock.

It may help at this stage to draw coloured lines along the ground surface for the various beds of rock, so that the outcrops of any one bed can be easily seen. To complete the section beneath the surface place a protractor on the point marking the *junction* of two beds on the line representing the ground surface and mark in the angle of dip below the horizontal (Fig. 8).

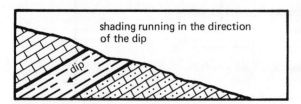

Fig. 9   Shading in the different rock types in a section.

Exercise 2

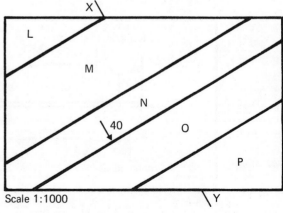

Scale 1:1000

Map showing outcrops of beds LMNOP

6

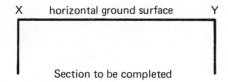

Copy the map and section shown above into your book, and answer the following questions.

Assume uniform dip and a horizontal surface

i   Name the oldest and youngest beds.
ii Mark on the map the strike of the beds.
iii Complete the section X—Y.
iv Find the true thickness and vertical thickness of beds M, N, O, and explain why the thicknesses of beds L and P cannot be found.

5 The effects of topography on outcrops

The land surface is, in fact, very rarely flat. Hills, valleys and ridges occur with slopes of varying steepness. These may be shown by contours or spot heights, but on simpler problem maps the relief is often indicated only on the outline section and by the direction in which rivers flow.

Where the beds are *horizontal* the oldest rocks will outcrop in the valleys and the youngest on the hill tops. The same bed will outcrop at the same height on all parts of the map, so that the pattern of the outcrops follows the contours. Figure 10 shows the outcrops of horizontal beds in an area of hills and valleys.

Where the beds are *vertical* the pattern of the outcrops is not affected by the topography and the beds will outcrop in straight lines across the map as shown in Fig. 11.

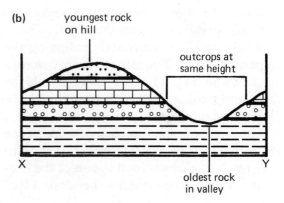

Fig. 10 **(a)** Geological map of the outcrops of horizontal beds in an area of hills and valleys.
(b) Section across the map in (a).

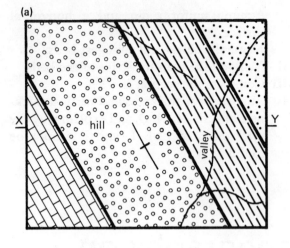

(a)

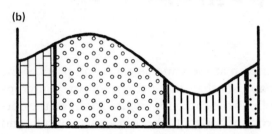

(b)

Fig. 11 **(a)** Map showing the same relief as Fig. 10 (a) with the pattern of outcrops produced by vertical beds. **(b)** Section showing vertical beds. It can be seen that whatever the relief is like the position of the outcrop on the surface will not change.

(a)

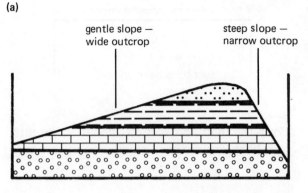

(b)

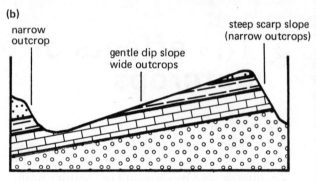

Fig. 12 **(a)** Horizontal beds. **(b)** Dipping beds outcropping across an escarpment and valley. An escarpment is a hill area with a gentle slope in the direction of dip and a steep scarp slope cutting across the beds to produce a series of narrow parallel outcrops.

More often the beds are dipping at an angle somewhere between vertical and horizontal, but two generalisations can be made from the above examples:

a When the dip is gentle the outcrops tend to follow the relief features.

b When the dip is steep the outcrops are almost straight whatever the relief.

Where beds are horizontal or dipping the slope of the land will alter the width of outcrops as shown in Fig. 12. This effect should be taken into account when comparing the widths of outcrops on maps.

The steepness and direction of slope of the land often vary even over a small area and this may result in changes in direction of the outcrops. Valleys cutting into the slope often produce V shaped outcrops. Whether the V points up or down the valley, depends on the direction and angle of dip of the rocks and the steepness of the slope of the land. Figure 13 shows three different relationships between dip and slope:

(a)

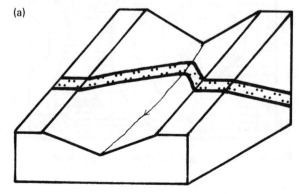

(b)

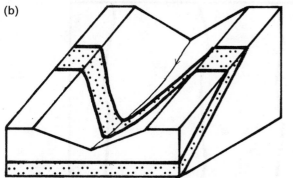

(c)

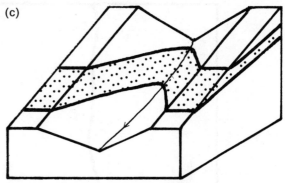

Fig. 13 Three different relationships between dip and slope. **(a)** Dip in opposite direction to slope, V points upstream. **(b)** Dip steeper than slope, V points downstream. **(c)** Slope steeper than dip, or beds horizontal, V points upstream (see outcrop pattern on Fig. 10).

Exercise 3

Using the three maps below, complete three separate sections to show the dip in each case. The topography is the same on all three maps —a simple ridge and valley. Draw in a dashed line to indicate the possible base of the oldest bed. The relative ages of the beds are the same in all three sections.

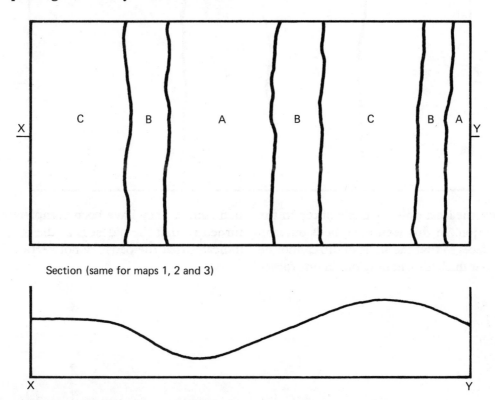

Section (same for maps 1, 2 and 3)

i The direction of the dip can easily be found as there are three points to join up between beds A and B and beds B and C.

9

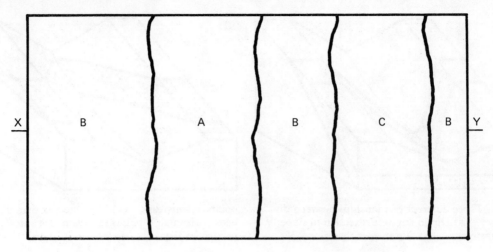

ii Copy this sentence into your book beneath the second section, completing the gaps:

The beds in this section are _____ as the outcrops are at the same _____ on either side of the hill and the valley.

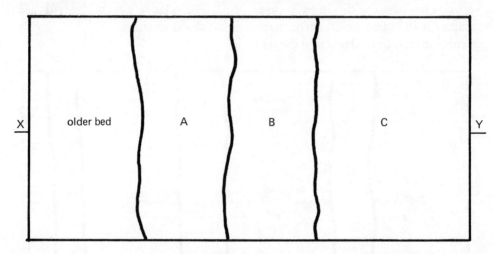

iii Show the least possible angle of dip in the third section. In this section no beds outcrop in two places so that the angle of dip cannot be found, but the beds can only dip in one direction (unless they have been completely overturned so that the oldest is at the top). What indicates that the beds are not vertical?

6 Inliers and outliers

Isolated areas of a particular rock can sometimes be seen on a map completely surrounded by rocks of different age.

An *inlier* is an area of older rock completely surrounded by younger rocks.

An *outlier* is an area of younger rock completely surrounded by older rocks.

The junction between the rocks of the inlier or outlier and the surrounding beds may be conformably bedded, an unconformity (see chapter 7), or a fault (see chapter 8). Inliers most often occur in valleys or where upfolds have been eroded exposing the older rocks in the middle. Outliers commonly form hills where the younger beds have not been eroded away. Figure 14 shows a map and section where an inlier and an outlier have been formed in gently dipping conformably bedded strata.

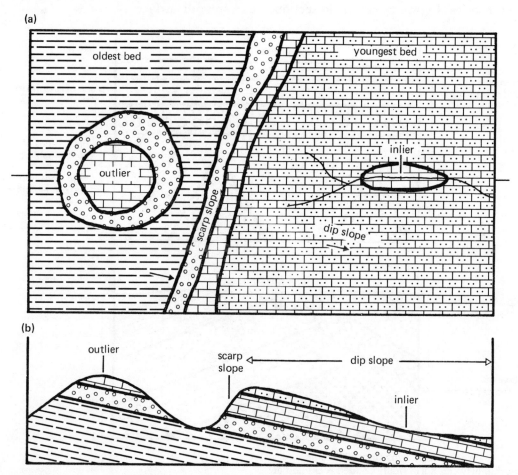

(a)

oldest bed   youngest bed

outlier   inlier

scarp slope

dip slope

(b)

outlier   scarp slope   dip slope

inlier

Fig. 14   An escarpment with a steep scarp slope cutting across the beds and a dip slope formed mainly of the youngest bed sloping down more gently in the direction of dip. Turn back to Fig. 10 (a) and find the outlier.

7 Unconformities

Unconformities occur where there has been a *break in deposition* between one series of beds and another. The older series will have been eroded during this break in deposition and this may give rise to a *conglomerate* at the base of the new series, formed from pebbles eroded from the older rocks and cemented together. The beds of the older series may also have been folded (see chapter 8) or tilted, and further tilting may have occurred since the formation of the newer series. Figure 15 shows, in section, two possible unconformable relationships. Unconformities can be recognised on maps where one bed cuts across a number of older beds (Fig. 16). The two sets of beds will normally have different directions and angles of dip or the younger beds may be horizontal.

When drawing a section across an unconformity it is best to draw in the younger beds first, then fill in the older ones beneath the plane of unconformity which should be taken as a smooth surface.

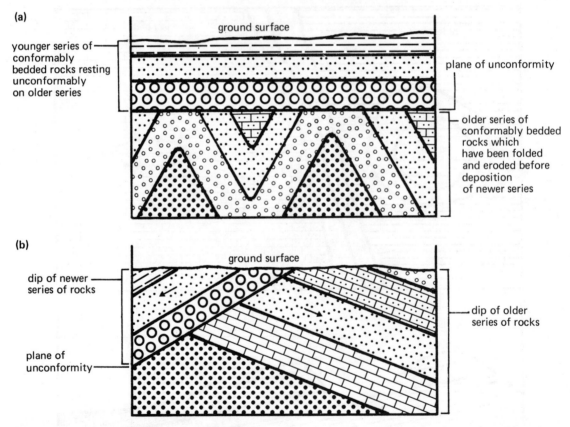

Fig. 15 **(a)** A section showing unconformable relationship between horizontally bedded rocks and an older series of folded rocks. **(b)** A section showing the unconformable relationship between two series of rocks with different dips.

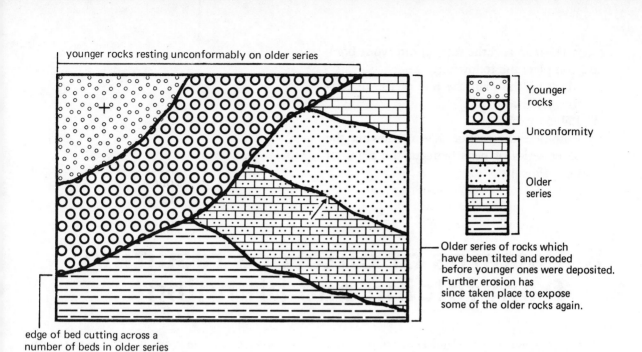

younger rocks resting unconformably on older series

Younger rocks

Unconformity

Older series

Older series of rocks which have been tilted and eroded before younger ones were deposited. Further erosion has since taken place to expose some of the older rocks again.

edge of bed cutting across a number of beds in older series

Fig. 16 Map showing younger rocks resting unconformably on an older series.

Exercise 4

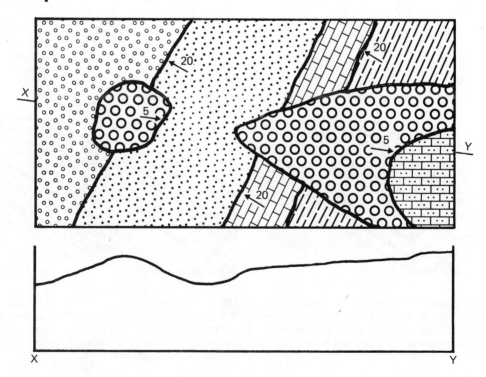

Copy the map and the section into your book and complete the following:

i   Mark on the map the line of the unconformity (two parts).
ii Label the outlier.
iii Complete the section X—Y showing the older series of beds beneath the plane of unconformity.

8 Patterns of outcrops resulting from folding and plunging

Rock strata may be folded as a result of pressure usually associated with earth movements. A simple upfold is called an *anticline* and a simple downfold a *syncline*. Dipping strata may be one side of a fold. Figure 17 shows block diagrams of an anticline and a syncline after erosion.

A *symmetrical fold* is one which has the same angle of dip on both sides or *limbs*. If one limb dips more steeply than the other the fold is *asymmetrical* and the outcrops of the steeply dipping limb will be narrower. The two limbs meet at the *fold axis* and a line can be marked in along the surface to show this. A line can

Oldest rock
exposed in
middle

(a)

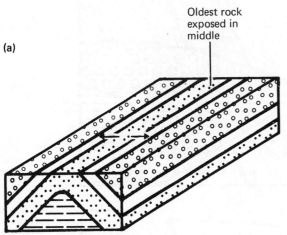

Youngest rock
preserved
in middle

(b)

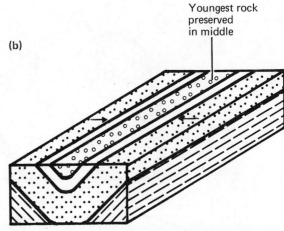

Fig. 17   **(a)** A simple symmetrical anticline. There are two parallel sets of outcrops with reversed orders. They become younger outwards and the dip is away from the middle—towards the younger rocks.

(b) A simple symmetrical syncline. Here again there are two parallel sets of outcrops, but the youngest rock is preserved in the middle and they become older outwards. The dip is towards the middle—once again towards the younger rocks.

also be drawn in on the section through the position of the axis for each bed, to show the *axial plane* (the surface at which the two limbs meet). The axial plane is not always vertical. Figure 18 shows the terminology of a fold.

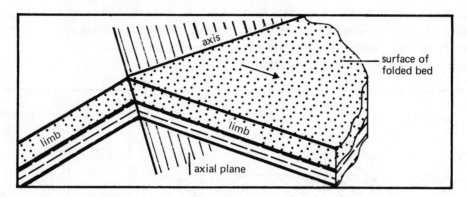

Fig. 18   The terminology of a fold.

Exercise 5

Copy the block diagram into your book, complete the pattern of outcrops and name the type of fold. Mark in the dip and label the youngest rocks and the lines shown.

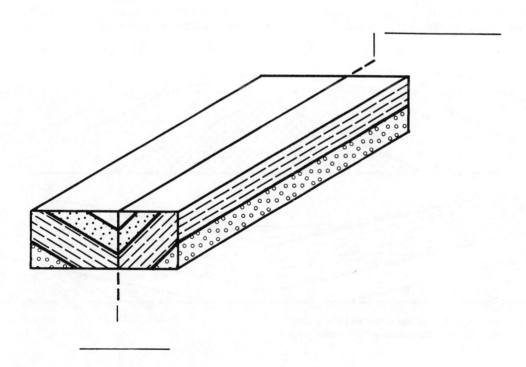

The pattern of outcrops shown by eroded folds can give a guide to structures in an area, but they will be more irregular where there are hills and valleys. An anticline will originally form a hill and a syncline a valley. However the top of an anticline is often weakened by stretching and the bottom of a syncline strengthened by compression due to the folding. An anticline may, therefore, be eroded to form a valley and a syncline may eventually form a hill. Figure 19 shows three sections through folded strata which illustrate how erosion may alter the relief; but note that the direction of dip and relative ages of the rocks still indicate an anticline and a syncline. Where some beds of rock are harder than others, erosion of an anticline may produce inward facing scarps and erosion of a syncline outward facing scarps (Fig. 20).

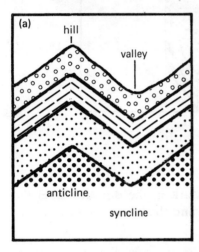
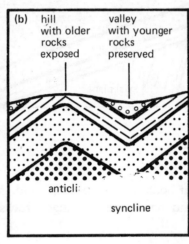
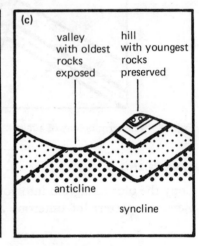

Fig. 19   Three sections through folded strata showing how erosion may alter the relief.

(a) Original fold.   **(b)** Reduced relief after some erosion.   **(c)** Inversion of relief.

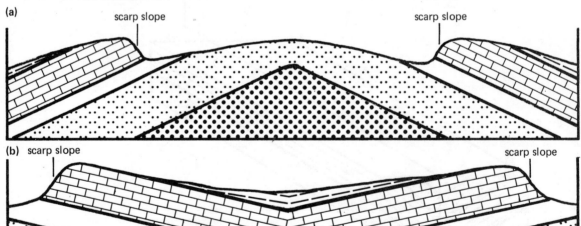

Fig. 20   Erosion of **(a)** an anticline, and **(b)** a syncline. In both cases the limestone is relatively hard and forms escarpments.

Exercise 6

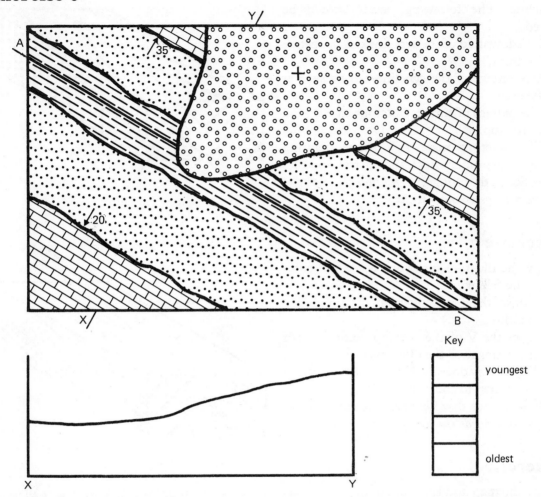

Key

youngest

oldest

Study the map and copy the section into your book and complete the following:

i   Show the rocks in order of age in a key.
ii   What structure is shown by the three oldest rocks?
iii What does the line A—B show?
iv   Comment on the relationship between the youngest bed and the three older beds.
v   Complete the section X—Y.
vi   What is the true thickness of the sandstone (it should be the same thickness on both parts of your section)?
vii Why is the outcrop of the sandstone narrower in the north of the map?

In the folds so far described the axes have been horizontal, but in some folds the axis and thus the whole fold may be tilted when the fold is said to *plunge*. A plunging anticline is shown in Fig. 21. After erosion plunging folds also show characteristic patterns of outcrops.

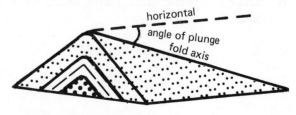

Fig. 21 A plunging anticline.

17

Figure 22 shows a block diagram of a plunging syncline. The following points should be noted:

a The downfold of the syncline can be seen at the front.

b The plunge of the syncline can be seen at the side.

c The pattern of surface outcrops can be seen on the top.

The outcrops make a V shape pointing towards the older rocks and in the opposite direction to the plunge, i.e. *the plunge (or pitch) is towards the younger rocks.*

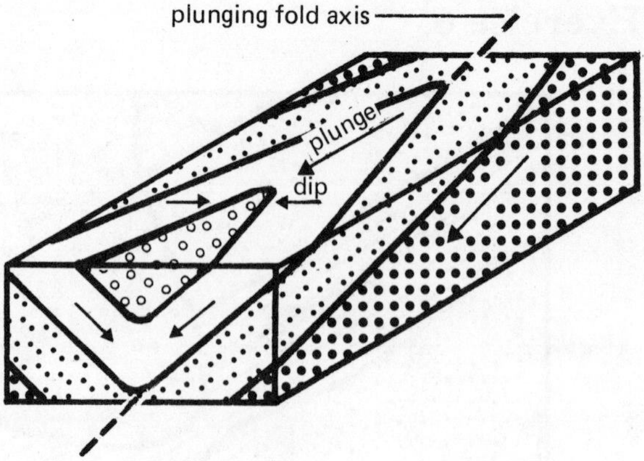

Fig. 22 A plunging syncline.

Exercise 7

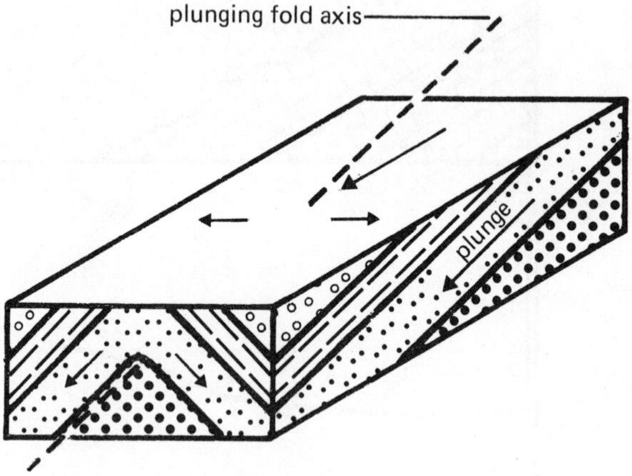

Copy the diagram into your book and complete the following:

i Finish the block diagram of a plunging anticline.

ii Does the V of the outcrop point towards the younger or the older rocks?

iii Which way does the V of the outcrop point in relation to the plunge?

You will see that here also the rule that *the plunge is towards the younger rocks* applies.

Exercise 8

Copy the map and the section into your book and complete the following:

i Finish the key showing the relative ages of the rocks.

ii Complete the section and show the oldest rocks.

iii On the map mark in two fold axes and show the direction of plunge along these.

iv Name the two main features along their axes.

v Suggest what events have taken place in the area since the rocks were first deposited conformably, to produce the present relief and outcrop patterns.

vi What name is given to the geomorphological features E, F, G and H shown on the section?

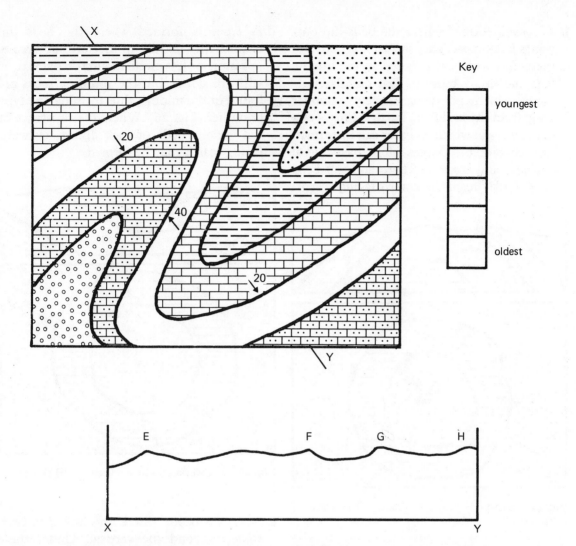

Other folded structures which can produce characteristic outcrop patterns are shown below with sections and maps.

a A *monocline* is a step-like bend in the beds which are mainly horizontal, but dip locally in one direction and then straighten out again (Fig. 23).

(a)

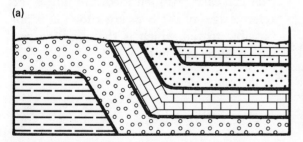

(b)

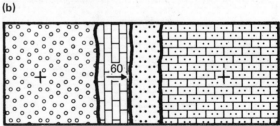

Fig. 23 (**a**) Section through a monocline.
(**b**) Outcrop pattern of (a).

b A *dome* is formed where the beds dip outwards in all directions, so that after erosion the oldest rocks are exposed in the middle surrounded by more or less circular outcrops which get progressively younger away from the middle (Fig. 24). This is one of the ways in which an inlier can be formed (see chapter 6). Domes often have radial drainage and erosion of hard and soft beds of rock may result in inward facing scarps.

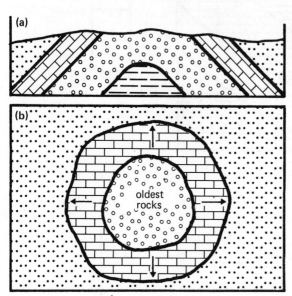

Fig. 24 **(a)** Section through a dome. **(b)** Outcrop pattern of (a).

c A *pericline* can be regarded as an elongated dome or an anticline plunging at both ends. After erosion it gives rise to oval outcrops with the oldest rocks in the middle (Fig. 25).

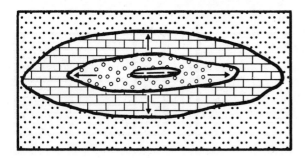

Fig. 25 Outcrop pattern of a pericline.

d A *basin* is formed where the beds dip towards the middle so that erosion results in a more or less circular outcrop pattern, as with the dome, but the youngest rocks are found in the middle, thus forming one type of outlier (Fig. 26). Where the resistance of the various beds of rock differs, outward facing scarps may be formed.

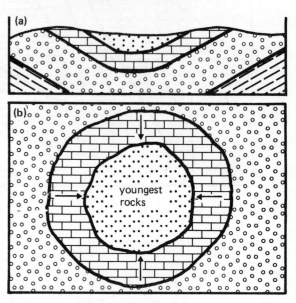

Fig. 26 **(a)** Section through a basin. **(b)** Outcrop pattern of (a).

e *Overfolds* occur where one limb has been folded beyond the vertical. Under these conditions the rule that the dip is towards the younger rock does not apply (Fig. 27). The repetition of the rock outcrops is typical of an anticline and a syncline and the fold axes can easily be identified, but unless the relative ages of the beds are known it is not possible to tell which is the anticline and which is the syncline.

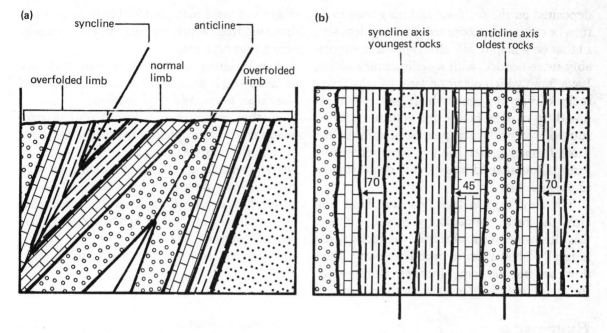

Fig. 27 **(a)** Section through an overfold.
(b) Outcrop pattern of (a).

9 Igneous intrusions and extrusions

Intrusive rocks are formed from *magma* which has cooled and solidified beneath the surface. They have *intruded* into already existing rocks along lines of weakness, such as faults (see chapter 10) and joints. Magma is molten (melted) rock formed deep down beneath the earth's surface. It may reach the surface through *volcanoes* or *fissures* (a line of small volcanic cones) when it forms a *lava flow*. These igneous rocks formed at the surface are *extrusive rocks* and are also called *volcanic rocks*.

Both groups of rocks are found at the surface, but for intrusive rocks to outcrop, the rocks which were above them must have been eroded. Extrusive rocks could also have been buried by later sediments and uncovered again by erosion. Both groups form characteristic features and the commonest of these, i.e. those shown on problem maps, are described below. Igneous rocks are usually hard and resistant to erosion so they often stick up above the general level of the surrounding sedimentary rocks.

a Volcanic rocks
By far the most important extrusive features are *basalt lava flows*. Basalt is a dark coloured, fine grained, basic igneous rock. It forms extensive lava flows which spread out and cover pre-existing strata. If extruded under the sea, the lava will cover the sediments already

deposited on the sea floor and may then in its turn be covered by more sedimentary deposits, so that on a map it will often appear conformably inter-bedded with a sedimentary series. Lava flows which occur on land usually cover rocks of varying ages, which have been exposed by erosion before the volcanic activity started, so the lava appears unconformable. Lava formed under the sea may sometimes form pillow shaped masses as it cools, known as *pillow lavas*.

The heat from a lava flow will bake and alter the rocks immediately beneath it (but this thin zone of metamorphosed rocks is not normally shown on a problem map). Lava flows are not the same thickness throughout; they tend to thin out away from their source of origin so that it may be possible to *suggest* the direction from which the lava came, if its thickness varies on a map.

A rule about *lava flows* is that they *are always younger than the rocks beneath them and older than the rocks overlying them.*

Further evidence of volcanic activity may be indicated on problem maps by the presence of other rocks of volcanic origin such as *rhyolite* (an acid igneous rock which is viscous and does not flow far from the point of extrusion), and *pyroclastic rocks* (formed from solid material thrown out by the volcano—the coarser material forming *agglomerate* or *volcanic breccia*, and the finer ash forming *tuff*.)

Exercise 9

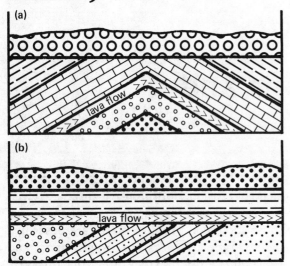

Complete the following in your book:

i Describe for each of the above sections the order of geological events which have resulted in the basalt lava flows being in their present positions. State whether the lava flows were submarine or terrestrial (formed on land).

Fig. 28 Block diagram showing a vertical dyke intruded into sedimentary rocks.

b Intrusive rocks

Intrusive rocks can be divided into two groups: *1.* those which form *minor intrusions and 2.* those which form *major intrusions.*

1. minor intrusions are small bodies of igneous rock intruded into spaces in the existing rocks such as along bedding planes, faults and joints. They are usually thin sheet-like bodies which may extend laterally for great distances. Where the igneous intrusion is more or less *vertical* it forms a *dyke* which cuts across the beds of rock and is therefore *discordant*. Figure 28 shows a dyke—note that the surface outcrop is straight, because the dyke is vertical, and that it cuts across other outcrops.

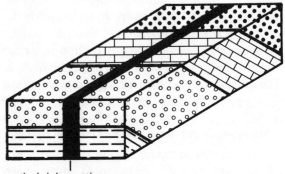

vertical dyke cutting across beds of sedimentary rock

Where the intrusion is more or less *horizontal* it forms a *sill* which is *concordant*, as it lies between the beds of rock. Some sills are described as *transgressive* because they pass from one bedding plane to another (Fig. 29).

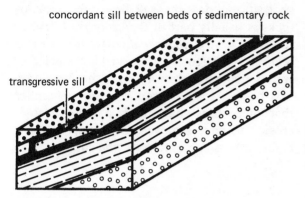

concordant sill between beds of sedimentary rock

transgressive sill

Fig. 29 Block diagram showing the difference between a concordant sill and a transgressive sill.

On maps, sills outcrop parallel to, and between, the beds of sedimentary rocks and, unless the land surface is flat (Fig. 29), their outcrops will bend, following the topographical features (compare Fig. 10). Sills may be confused with interbedded lava flows; the main differences are as follows:

a Sills are younger than the rocks above them as they were intruded after the surrounding rocks were formed.

b The rocks immediately above and below the sill will have been metamorphosed by the heat from the intrusion.

c Sills are formed by slightly coarser grained rocks than lavas (see below). The rock type will normally be the main way of distinguishing between sills and lava flows on problem maps.

Basalt should normally be regarded as an extrusive rock (although it may also form sills and dykes). The magma of minor intrusions has cooled relatively quickly compared with that forming major intrusions because it loses heat more rapidly to the surrounding rock (called the *country rock*), so that the crystals do not have time to grow large. On the other hand the magma cools more slowly than when it is extruded so these intrusive rocks are not so fine grained as extrusive ones. Rocks which commonly form minor igneous intrusions are *dolerite, microgranite, aplite, quartz porphyry,* and *pegmatite.* The medium grained rocks which form most of the minor intrusions are often classified as *hypabyssal*—ones formed at shallow depth and thus cooled more quickly than rocks formed deep down *(plutonic rocks).* This classification is not strictly correct as it is the size and thickness of the magma intrusion that is most important in determining the rate of cooling and thus the crystal size. Many minor igneous intrusions occur in association with major ones, e.g. microgranite dykes with granite bosses (see below).

Sills and dykes may occur in clusters forming such features as *linear* or *radial dyke swarms,* or *ring dykes.* These names describe the patterns of the outcrops. Other minor intrusive forms include *irregular veins* and *cylindrical plugs* (magma solidified in the neck of a volcano).

2. Major intrusions are formed by large accumulations of magma in the earth's crust. When exposed at the surface by erosion their outcrops may be many miles across, hence only the smaller types are shown on problem maps. *Granite* (a coarse grained acid igneous rock) is the most common rock forming major intrusions. It is normally shown forming an approximately circular outcrop called a *boss* (if the outcrop is irregular it is called a *stock*). Such outcrops may be the upper part of a much larger intrusion. They may be associated with folding and may occur in the crest of an anticline. The edges of these intrusions are steep, cutting across the country rock. The heat from the intrusion will have thermally metamorphosed the country rock in contact with it, forming a *metamorphic aureole* (see chapter 10 on metamorphic rocks, below). The edge of the aureole is usually shown on a problem map by a dashed line (Fig. 30).

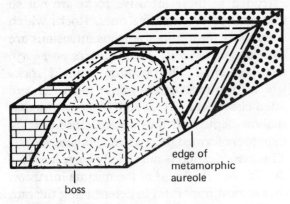

Fig. 30 A granite boss intruded into dipping sedimentary rocks.

In the key to a geological map intrusive rocks are placed in a separate column and their age, if known, is given in the margin. Extrusive rocks are placed with the sedimentary rocks as they are interstratified and their age is known. However, on problem maps, to suggest the possible age limits for igneous rocks is usually part of the exercise. *Intrusive rocks are always younger than the surrounding country rocks* (so it can never be correct to start a geological history with the formation of a boss). It is also important to remember that there must have been a great deal of erosion to have removed the overlying rocks before a boss can outcrop at the surface.

Exercise 10

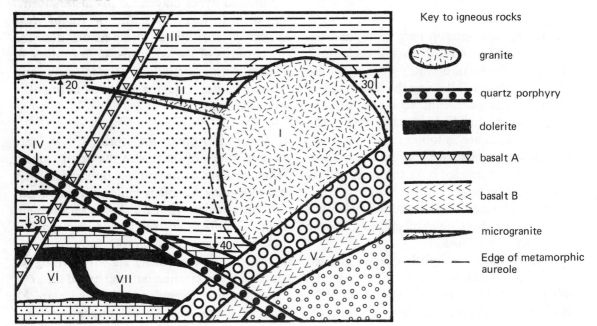

Complete the following questions in your book:

i   Name the igneous features numbered I to VII on the map.

ii   List their relative ages, giving your evidence.

iii List the sedimentary rocks in order of age.

iv   What two structures do the outcrops of the sedimentary rocks reveal?

10 Metamorphic rocks

As has been stated above metamorphic rocks may result from igneous intrusions when the country rock is altered by the heat. New rocks formed within the metamorphic aureole by this *thermal metamorphism* are not often shown on problem maps, but it is possible to suggest what alterations will have taken place. In general the metamorphosed country rocks will be harder and more resistant to erosion than the unaffected rocks. Some important changes which take place when certain rocks are thermally metamorphosed are as follows: Pure limestone re-crystallizes to form *marble* and pure sandstone to form *quartzite*. In rocks such as clay and shale (argillaceous rocks) new minerals may be formed giving rise to *spotted slates* and *hornfels*. The latter forms at a higher temperature and is thus found closer to the intrusion.

Exercise 11

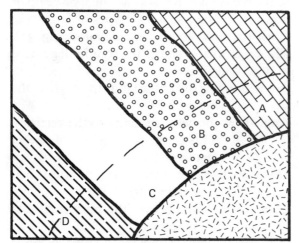

Study the map and complete the following:

What rocks would you expect to find at A, B, C and D?

Other metamorphic rocks which may be shown on problem maps are those which result from *regional metamorphism*. They have been greatly altered by varying amounts of heat and pressure and include *schists* and *gneisses*. These rocks usually occur on problem maps as the oldest rocks, e.g. exposed in the middle of an anticline, and they represent rocks of a much older series than the sedimentary rocks shown above them which have not been metamorphosed. *Slates*, which result from the effect of pressure on argillaceous rocks, may occur interbedded with sandstones and limestones which have not been much altered by pressure. Slates would normally belong to the older series of rocks shown on the map, even if they are not the oldest rock, as they indicate that these rocks have experienced some low grade metamorphism.

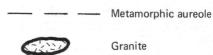

——— — ——— Metamorphic aureole

Granite

11 The effects of faults on outcrops

A fault is a fracture in the earth's crust, along which the rocks on one side have moved relative to those on the other side. The movement may be vertical, horizontal, or both combined. Faults in which movement has been vertical or horizontal are described below. Faults are shown on maps by a heavy line sometimes marked F—F on problem maps. Where movement along the fault has displaced the rocks vertically the *downthrown side* (the side which has gone down *relative* to the other, whichever side actually moved) is indicated by 'ticks' (Fig. 31), but on many problem maps this is one of the things to be worked out.

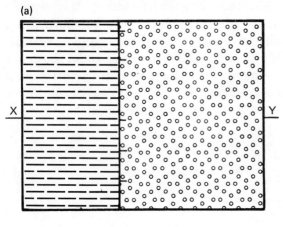

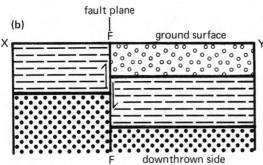

Fig. 31 **(a)** Map of faulted horizontally bedded rock. **(b)** Section of (a).   F—F = fault.

On the simpler problem maps faults are shown as straight or almost straight lines which show that the *fault plane* (the surface along which the movement has taken place) is vertical or very steeply inclined and therefore not, or only slightly, affected by relief. In a section a fault should be drawn as a vertical line if its outcrop is straight. If the outcrop is almost straight draw the fault as a steeply inclined straight line (to show a normal fault in the absence of any evidence to the contrary—see below). Figure 31 shows in map and section the very simple case of horizontal strata cut by a vertical fault plane along which there has been vertical movement. Note the symbol used in the section to indicate the relative movements on the two sides of the fault. The younger rocks are found on the downthrown side.

In fact the fault plane is often steeply inclined rather than vertical and the movement of the rocks on either side can take place up or down it, giving rise to two types of faults: *1. normal faults* and *2. reverse faults*. (Fig. 32). Figure 32(a) also shows the terminology used in describing faults:

Angle d = the dip of the fault (the angle to the horizontal).

Angle h = the hade (the angle to the vertical). This term is going out of use in favour of dip.

Distance T = the throw (the vertical displacement).

Distance H = the heave (the horizontal displacement).

A reverse fault with a small angle of dip is called a *thrust* and on a map the fault line will not be straight, as its outcrop is influenced by relief in the same way as outcrops of gently dipping strata. If a bore hole was sunk from point B in Figs. 32(a) and 32(b) the shaded bed would be missed in the case of the normal fault and encountered twice in the case of the reverse fault.

(a)

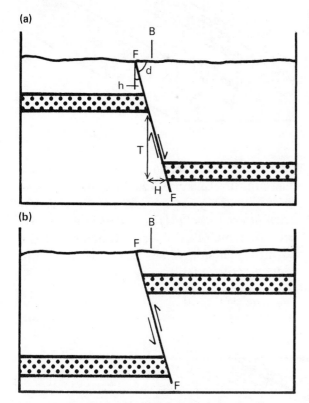

(b)

Fig. 32 **(a)** Section through normal fault.
(b) Section through reverse fault.

When a fault is first formed the land on the upthrown side will be higher, but as it is more liable to erosion the land surface is gradually levelled off. However, a section may reveal a change of slope at the point where the fault reaches the surface, especially if more resistant rocks have been brought against easily eroded ones. The steep slope which results, is called the *fault line scarp* (Fig. 33). A fault scarp is the immediate result of faulting before erosion. Faults may occur in any direction through the rocks and they can be classified by their relationship to the dip of the rocks which they affect:

a *Strike faults* faults which run in the direc-

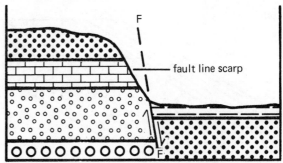

Fig. 33 Section through a fault line scarp.

tion of the strike.
b *Dip faults* faults which run in the direction of the dip of the rocks (this term has nothing to do with the dip of the fault plane).
c *Oblique faults* faults which cut obliquely across the strike, i.e. those which are not parallel to either the dip or the strike.

The effects of the first two types on outcrops are described below and should give enough understanding for the interpretation of any faults, where there has been vertical movement.
a *Strike faults* can affect the patterns of outcrops in two ways depending on whether the downthrow is in the direction of dip of the rocks or opposite to it.

Exercise 12

Block diagram A shows the case where the rocks are downthrown in the direction of dip. Block diagram B will show the case where the rocks are downthrown against the dip.
Copy diagram B into your book and complete the following:
i Complete the surface outcrops in diagram B.
ii What has happened to bed X in A and in B?
iii On which side of the fault do the younger rocks occur in each case (consider only the rocks brought in contact with each other across the fault)?

27

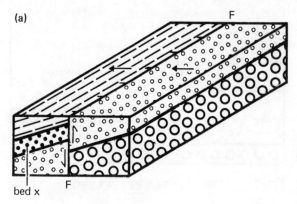

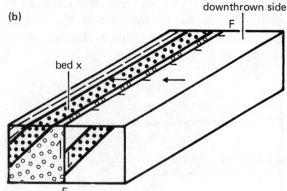

The answers to Exercise 12 will have shown that strike faults can cause repetition of outcrops when downthrown against the dip, and they can cause beds to be cut out at the surface when downthrown with the dip. This is true whether they are normal or reverse faults, as can be seen by altering the inclination of the fault plane in the sections shown in the block diagrams in Exercise 12. The answers also show that the *younger rocks are found on the downthrown side of the fault*. This is an important rule which can be used to determine the downthrown side where two beds come into contact across a fault.

Exercise 13

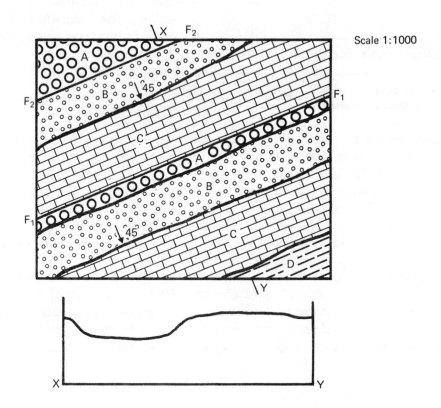

Study the map, copy the section into your book and complete the following:

i Complete the section X—Y.

ii Measure the thickness of bed B and make sure it is shown correctly between the two faults in the section.

iii Calculate the throw of the fault F_1 (extend the top of bed B until it meets the fault from both sides; the vertical displacement of the bed can then be measured).

In the section in Exercise 13 the land between the two faults has slipped down to form a *rift valley*. When the land between two faults has moved up to form a hill the feature is known as a *horst*. Where a series of parallel faults occur all downthrown in the same direction they are known as *step faults* and they may result in successively younger strata outcropping across an area. These terms apply to any faults where such vertical movements take place.

Exercise 14

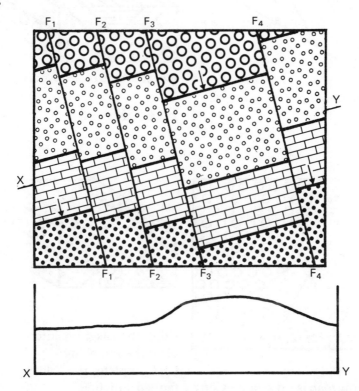

i Study the map and copy the section into your book. Complete sketch section X—Y to show the relationship of faults and beds of rocks (the thickness of the beds cannot be determined, but they should be related to the widths of their outcrops).

ii Label the two structural features shown.

b *Dip faults* also produce characteristic outcrop patterns in tilted and folded strata. The block diagram (Fig. 34) shows how outcrops are displaced by vertical movement along a dip fault. All the outcrops on the upthrown side of the fault have been shifted in the direction of dip of the rocks. A similar effect can be produced by a tear fault in which movement is horizontal (see below). The apparent lateral displacement is the result of erosion of the

29

tilted strata to approximately the same level on both sides of the fault. Once again it can be seen that the younger rocks on the downthrown side have been brought against older rocks on the upthrown side.

that the youngest rock is preserved in the middle of the syncline on this side.

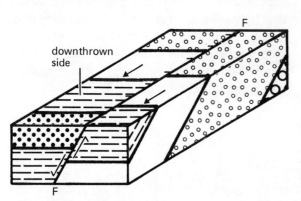

Fig. 34 Block diagram showing the effects of a dip fault on dipping strata.

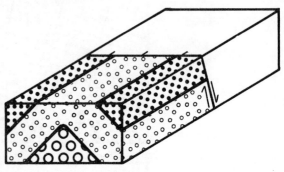

Fig. 35 A dip faulted syncline.

Exercise 15

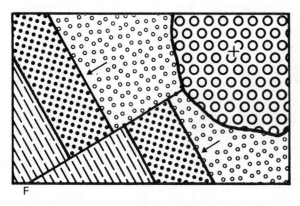

The fault F—F on the map is a dip fault. Copy the map into your book and complete the following:

i Comment on the age of the fault.

ii Which is the oldest rock shown on the map?

iii Mark on the map the downthrown side of the fault.

Figure 35 shows a dip faulted syncline. It can be seen that there is a wide outcrop of younger rocks on the downthrown side and

Exercise 16

Copy the diagram into your book and complete the following:

i Complete the block diagram of a dip faulted anticline.

ii On which side of the fault do more older rocks outcrop?

iii State this in terms of the younger rocks.

Plunging folds may also be faulted resulting in either repetition of part of the V shaped outcrops or in part of the V being cut out at the surface. Figure 36 shows a faulted plunging anticline downthrown in the same direction as the plunge which has resulted in part of the outcrops being cut out.

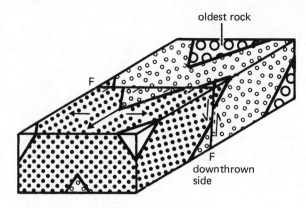

oldest rock

F

F

downthrown side

Fig. 36 A dip faulted plunging anticline.

Exercise 17

downthrown side

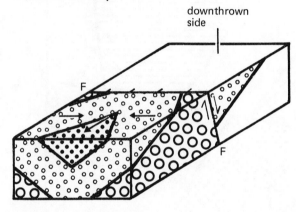

F

F

Copy the diagram into your book and complete the block diagram of a faulted plunging syncline downthrown in the opposite direction to the plunge (see Fig. 21). Note the repeat of the V shaped pattern of the outcrops. There are two other possibilities:

a A faulted plunging syncline downthrown in the direction of plunge.

b A faulted plunging anticline downthrown in the opposite direction to the plunge.

You may like to draw block diagrams of these also. These outcrop patterns, though complicated, can be interpreted by using the three basic rules:

a *The dip is towards the younger rocks.* This shows whether it is a syncline or an anticline (Fig. 17).

b *The plunge is towards the younger rocks.* This gives the direction of plunge (Fig. 22 and Exercises 7 and 8).

c *Younger rocks occur on the throwndown side of the fault* against older rocks on the upthrown side. This rule will always reveal the downthrown side where there has been vertical movement.

31

Exercise 18

The key given shows the beds in order of age.

Key

Copy the map, into your book, and complete the following:

i Label the geological structures shown on the map and mark the dip and plunge with arrows.

ii Show the downthrown side of the fault.

Tear faults are formed where there has been horizontal (sideways) movement along an almost vertical fault plane. The effect is to displace the surface outcrops (as with dip faults in the case of simple tilted strata). The difference between the effects of tear faults and dip faults on outcrop patterns is that *tear faults displace all outcrops an equal amount in the same direction whatever their dip,* whereas the displacement of outcrops by a dip fault depends on the amount and direction of dip (see above). Thus a tear fault will displace a vertical dyke along with the dipping strata it cuts through as shown in Fig. 37. Where vertical movement takes place the dyke outcrop is not displaced. Note that vertical movement appears to have occurred in the side section of the block diagram due to the horizontal shift of the tilted strata.

In tear faulted folds all the outcrops are moved in the same direction and the beds of the two limbs of the fold remain the same distance apart on both sides of the fault (Fig. 38).

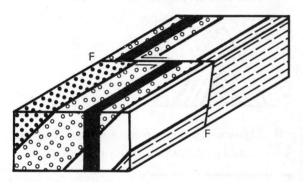

Fig. 37 Tear fault across dyke in dipping strata.

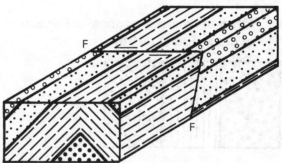

Fig. 38 Tear faulted anticline.

Exercise 19

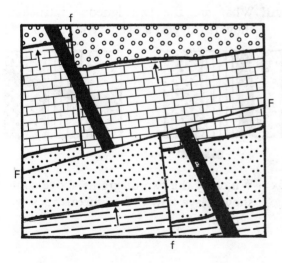

Copy the map into your book and complete the following:
i   List the sedimentary rocks in order of age.
ii Which is the youngest rock shown on the map and what feature does it form?
iii Assuming that all the rocks were present before any faulting occurred, which of the two faults is the older?
iv What type of fault is f—f?
v   What type of fault is F—F?
vi Explain how it is possible to tell the difference between the two faults.
vii Mark the downthrown side of the fault(s) where there has been vertical movement and indicate the direction of horizontal movement along the faults if it has occurred.

33

Exercise 20

This series of short revision questions is based on all the topics covered so far. Assume the land surface is flat. All diagrams except xii are simple maps of surface outcrops. The following symbols apply throughout. Complete the

 oldest rock   youngest rock

following questions in your book, copying the diagrams where necessary.

i   Draw in an arrow to show the direction of dip and state the rule.

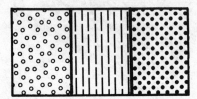

ii   If the beds are as thick as in i above, why is their width of outcrop different?

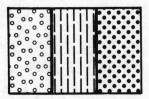

iii   Draw a section to work out the thickness of the middle bed.

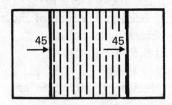

iv   Define an outlier and an inlier.

v   What structure is shown here? Mark in the dip.

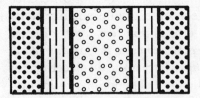

vi   What is this structure? Mark in the dip etc. with arrows and label them.

vii   Draw a simple map to show how you can identify an unconformity on a map.

viii   What are the two igneous intrusions shown? Suggest a likely type of rock.

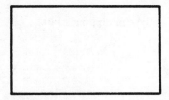

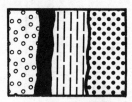 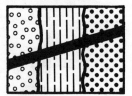   intrusive rock

ix   Which of the two diagrams above could show a lava flow if the igneous rock was extrusive?

x What igneous feature is shown here? What is represented by the dashed line? What rock would you expect to find at X and at Y.

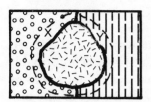

xi Which side of the fault is downthrown? What sort of a fault is it?

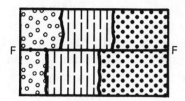

xii Complete the block diagram.

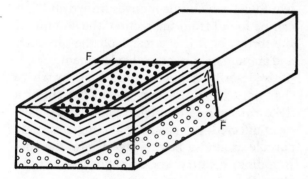

xiii Draw a section to show why the outcrops are repeated. Which side is downthrown? What type of fault is it?

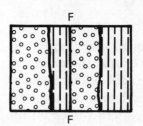

xiv Assuming all the rocks were present before faulting, name the two types of fault and explain the difference in the outcrop patterns.

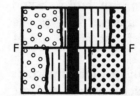 

12 The complete geological history

Most questions on problem maps expect the student to be able to work out all the structures shown and their relative ages of formation, so that an orderly account of the geological events leading to the outcrop pattern shown on the map can be written. The shortest form of the question will be 'write a geological history of the area shown on the map provided'. Other questions may give more guidance by asking definite questions about, for example, the ages of the rocks and the types of faults. Some examples of problem maps set by various examining boards will be found at the end of this book.

In writing a geological history it is important to follow a plan or the account will become muddled. It is probably helpful to note in rough the features such as unconformities, folds and faults as they are identified and to put them in order of occurrence. It is also helpful to number all the rocks shown in order of age, starting from the oldest. Only then are you ready to start writing the geological history. If there is a section to complete see that the order of rocks and the structures shown in it, correspond to those described in the account—this is a useful cross check.

1. Start the description by naming the oldest rock and those deposited conformably upon it. Describe where on the map they outcrop and the trend of the outcrops. From evidence of the type of rock and information such as which fossils occur in the rock, briefly describe the conditions of deposition.
2. What happened after this conformable series was deposited? Describe any tilting, folding, faulting or igneous activity. This series must at least have been uplifted and eroded before another series was deposited unconformably on top.
3. Describe, in a similar way to 1 and 2 above, the next series of rocks resting unconformably on the first and any later rocks if there is more than one unconformity. If the age of one particular rock (most likely an igneous intrusion) cannot definitely be determined discuss the various possibilities.
4. When describing tilting and folding give the direction and amount of dip, name the types of folds present, give the trend of the fold axes and state whether the folds are symmetrical or asymmetrical, (do the widths of outcrops vary as a result)?
5. Descriptions of faulting should include the direction of the fault, the type (dip, strike or tear fault), which side is downthrown and, if it can be found from the section, the amount of the throw, or if it is a tear fault the direction and amount of horizontal displacement.
6. In describing igneous intrusions say where they occur, the direction in which a sill or a dyke runs across the map, which rocks have been affected, the extent of the metamorphic aureole around a boss (suggest how the surrounding country rocks might have been altered).
7. Having described in detail all the events leading to the deposition of the most recent bed shown on the map, remember that the rocks are now above sea level and are being eroded. Describe how the rocks influence the relief of the area—look at the section, coastline and direction of streams for evidence of hills, valleys, headlands and bays, and relate them to the varying resistances of the beds. Does the relief affect the widths of outcrops of particular beds?
8. Give as far as possible (usually from the section) the thickness of the various beds and

point out where the width of outcrop varies as a result of the different thicknesses. Also show how variations in dip can cause beds of different thicknesses to have similar widths of outcrop.

9. Describe any special features shown on the map, e.g. quarries—suggest why a particular rock has been quarried. Is there any evidence of longshore drift along the coast or recent sea level changes? Are there any pot-holes or disappearing streams in limestone? Is there any sign of glaciation and can the direction of ice movement be determined?

A sample geological history and section for the map shown in Fig. 39 is given below. The following points should be noted about the section:

i The rocks above the unconformity are shown with a dip of less than seven degrees, as the section is not drawn along the dip.

ii The thickness of the limestone is shown on the section near A where it dips at 30 degrees, so the limestone of the other limb of the syncline (cut by the fault) has been shown with the same thickness.

iii The sandstone has been shown $\frac{2}{3}$ the thickness of the limestone, as its outcrop across the middle of the map is about $\frac{2}{3}$ as wide as that of the limestone (where both dip at 45 degrees).

Geological history of map

The oldest rock outcropping on this map is the mudstone. Sandstone, limestone and clay were deposited conformably on top of the mudstone—all four sediments being deposited under marine conditions. They outcrop in bands running almost ENE—WSW across all, except the southern part of the map where they are covered by later deposits. These four sediments were folded to form an anticline in the south and a syncline in the north with the trend of the fold axes ENE—WSW, parallel

to the present outcrops. Both folds are asymmetrical with their common limb dipping most steeply at 45 degrees to the NNW. The southern limb of the anticline has a dip of 20 degrees to the SSE and the northern limb of the syncline dips at 30 degrees in the same direction. Since folding these beds have been faulted, eroded and affected by igneous intrusions. Uplift and the start of erosion could have been associated with the folding. The dolerite dyke running east–west across the northern part of the map was certainly intruded before the western fault occurred as its continuity is broken by this tear fault. The position of the outcrop of the dyke would not have been altered by the eastern fault where there was vertical movement so that it is not possible to tell which occurred first, but it is likely that the two faults occurred at approximately the same time and the dyke is older than both faults.

The western fault is a tear fault along which the rocks on the western side have been displaced northwards by 1000 metres relative to those on the eastern side. The other fault is an oblique fault (which has had the same effect as a dip fault) downthrown to the west where the youngest rock of the series, clay, has been preserved in the middle of the syncline. From the section the throw of the fault appears to be a little over 100 metres.

The next event was the intrusion of the granite boss in the north west. Associated with this is the metamorphic aureole, and it can be seen from this that metamorphism occurred after the tear fault. Within the aureole the limestone, if pure, will have re-crystallized to form marble, the sandstone to form quartzite, and the clay will have been altered to form a rock such as hornfels.

After all these events there must have been a long period of erosion to expose the granite

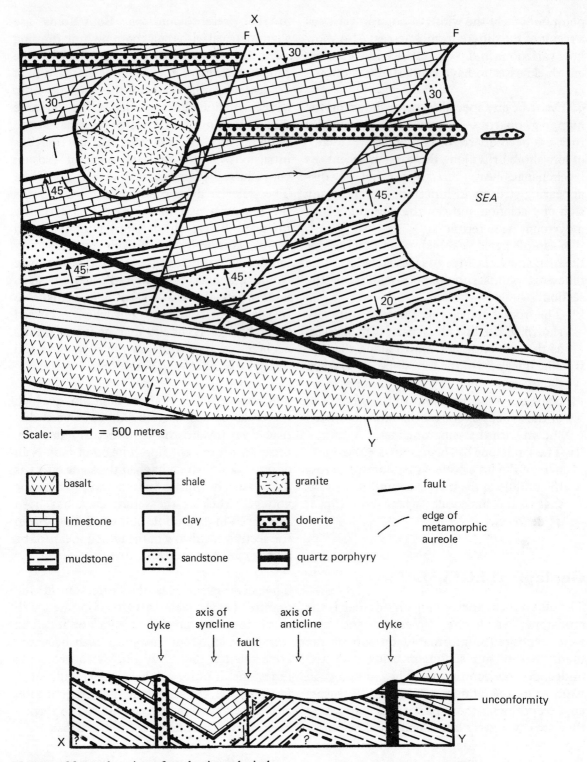

Scale: ┣━━━┫ = 500 metres

basalt	shale	granite	fault
limestone	clay	dolerite	edge of metamorphic aureole
mudstone	sandstone	quartz porphyry	

Map 39 Map and section referred to in geological history.

boss at the surface. Eventually the area must have been submerged again, when the shale was deposited unconformably on the older rocks. The two shale deposits which outcrop across the middle part of the map are separated by basalt, which would probably have been formed by a submarine lava flow. The shale, now remaining only in the south west of the map, was deposited on top of the lava flow and is probably older than the *quartz porphyry dyke* which cuts across the basalt and the lower shale as well as the older rocks (but as its relationship to the shale cannot be seen their relative ages cannot definitely be determined). This quartz porphyry dyke is therefore probably the last feature to be formed and either before or after its formation the area was again uplifted and tilted south, south west producing a dip of 7 degrees. Since then there has been further erosion removing the younger rocks to expose the older series over much of the area.

From both the section and the coastline it can be seen that the dolerite dyke is particularly resistant, forming a ridge inland and a headland at the coast. The quartz porphyry dyke is also resistant as can be seen in the section. The map shows streams draining from the granite boss which indicate that this is also higher land than the surrounding country rock. The lowest land and the largest bay are formed by the mudstone which occurs in the middle of the anticline—this is a soft rock which was probably further weakened by folding. The sandstone appears to be slightly more resistant than the limestone, as it has been slightly less eroded at the coast. In the south the edge of the basalt forms a steep slope above the clay.

The limestone has a true thickness of about 500 metres, the sandstone about 300 metres (from section), but the thickness of the mudstone and clay cannot be measured. The shale below the basalt is also about 200 metres thick and as the basalt has a much wider outcrop than the shale it must be thicker—about 500 metres. The effect of the dip on the widths of outcrops can be seen best by comparing the limestone outcrops dipping at 45 degrees and 30 degrees where the outcrops are 750 and 1000 metres respectively. The streams flow mainly along the clay after leaving the granite, as the clay is softer and impermeable, and forms the lowest land. The course of the stream flowing east to the coast has been influenced by the two faults, which it follows for short distances and near its mouth it has been turned east again by the dyke. The fact that the limestone has been more eroded south of the dyke than north of it, may be the result of the stream valley reaching the coast here or possibly that the dyke has protected the limestone north of it from erosion by the sea.

This geological history is rather long as it attempts to explain fully how the various conclusions were reached and to cover every minor point. It does however show how much information can be found by studying a geological map. After working through the following exercises you should be able to produce a detailed geological history for any such problem map.

Map 1 Study the geological map below.

Complete the following questions, in your book:
a Draw the geological section from X to Y
b Name the type of intrusion formed by the dolerite
c What is the oldest rock type shown on the map?

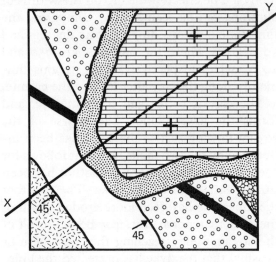

KEY

■	Dolerite	∘∘∘	Shale
	Volcanic Ash		Mudstone
	Fine-grained sandstone	↗45	Direction and amount of dip in degrees
	Coarse-grained sandstone	+	Horizontal strata
	Chalk		

Map 2 Study the partly completed geological map provided which shows the outcrops of horizontal strata. Part of the surface area is covered with vegetation and no rock exposures are visible. Copy the map in your book and complete the following:
a Complete the geological map of the area
b Using the symbols given in the key draw your own geological column showing the correct sequence of the beds (oldest at the base)
c Mark on your map with the letter X one place where a spring might be expected to occur.
d How many feet thick is the flagstone outcrop?

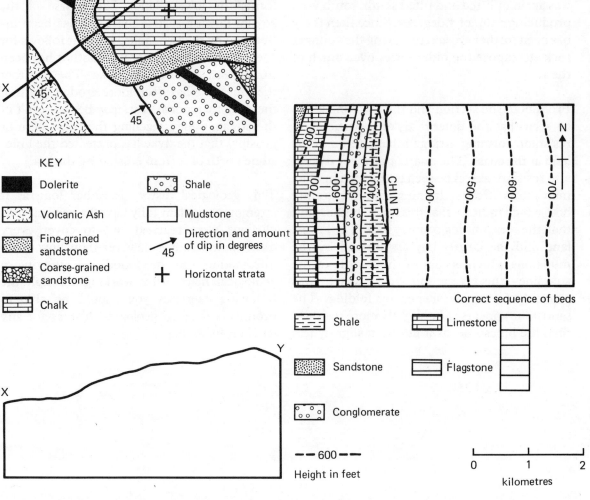

Correct sequence of beds

	Shale		Limestone
	Sandstone		Flagstone
∘∘∘	Conglomerate		

- - 600 - -
Height in feet

0   1   2
kilometres

Map 3 Carefully study the geological map given below.

a Copy the profile provided into your book and draw as complete a geological section as possible from X to Y

Complete the following statements correctly, in your book:

b The youngest formation shown on the map is . . . . ?

c The oldest formation shown on the map is . . . . . ?

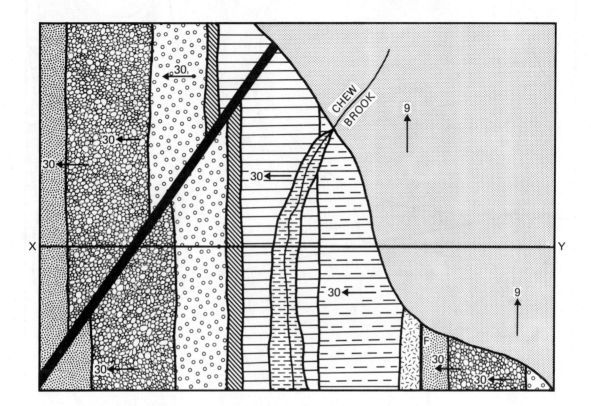

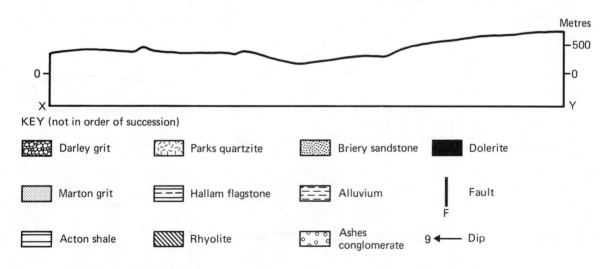

KEY (not in order of succession)

Darley grit		Parks quartzite		Briery sandstone		Dolerite
Marton grit		Hallam flagstone		Alluvium		Fault
Acton shale		Rhyolite		Ashes conglomerate		9 ← Dip

Map 4

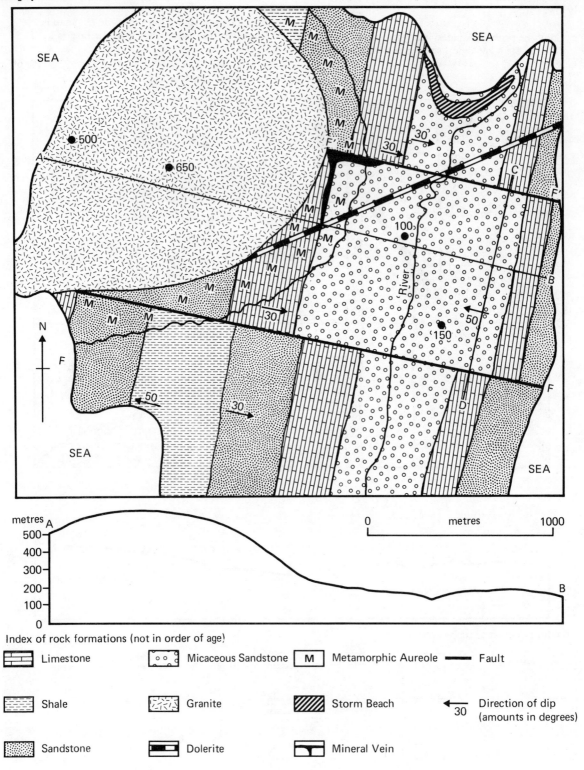

Index of rock formations (not in order of age)

Limestone	Micaceous Sandstone	M Metamorphic Aureole	Fault
Shale	Granite	Storm Beach	Direction of dip (amounts in degrees) 30
Sandstone	Dolerite	Mineral Vein	

a Complete the section A-B on the map in the profile provided

b *Sketch* a section along the line C-D

c Name the structure illustrated by C-D

d Describe the folding on the map under the following headings: *(i)* strike, *(ii)* dip of limbs, *(iii)* symmetry

e Put into geological sequence the following events (start with the oldest): intrusion of granite; laying down of sedimentary series; intrusion of dolerite; formation of mineral vein; folding of sedimentary series

f Describe the likely changes to be found in: *(i)* the sandstone, *(ii)* the limestone, *(iii)* the shale within the matamorphic aureole

g Account for two features of interest in the course followed by the river

h Describe briefly the effect of folding on the width of the limestone outcrops

Map 5

a Copy the profile into your book and complete the geological section from A to B

b Write an account of the geological history of the area shown on the map, including reference to geomorphological features related to contrasting rock types and to variations in widths of outcrop

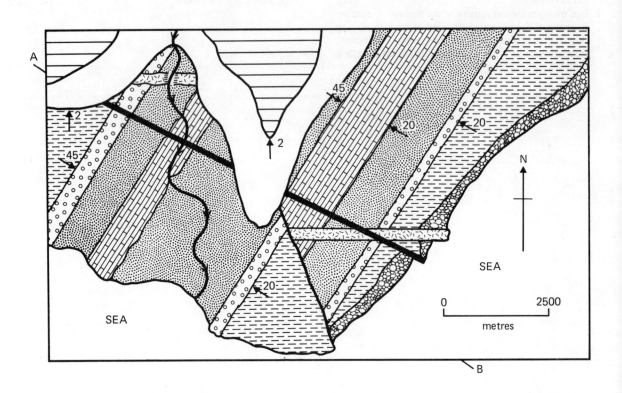

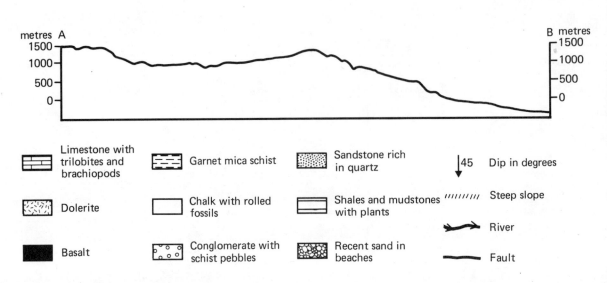

	Limestone with trilobites and brachiopods		Garnet mica schist		Sandstone rich in quartz	↓45	Dip in degrees
	Dolerite		Chalk with rolled fossils		Shales and mudstones with plants	/////////	Steep slope
	Basalt		Conglomerate with schist pebbles		Recent sand in beaches		River
							Fault